RESTAURANT DESIGN

GID餐厅设计作品集

RESTAURANT DESIGN PORTFOLIO OF GID

曾建龙 编著

江苏凤凰科学技术出版社·南京

图书在版编目（CIP）数据
GID餐厅设计作品集 / 曾建龙编著 . -- 南京：江苏凤凰科学技术出版社，2025. 3. -- ISBN 978-7-5713-4945-5
Ⅰ. TU247.3
中国国家版本馆CIP数据核字第2024JA9975号

GID餐厅设计作品集

编著	曾建龙
项目策划	凤凰空间 / 杜玉华
责任编辑	赵 研
责任设计编辑	蒋佳佳
特邀编辑	杜玉华
出版发行	江苏凤凰科学技术出版社
出版社地址	南京市湖南路1号A楼，邮编：210009
出版社网址	http://www.pspress.cn
总经销	天津凤凰空间文化传媒有限公司
总经销网址	http://www.ifengspace.cn
印刷	天津裕同印刷有限公司
开本	965 mm×1270 mm 1／16
印张	14
插页	18
字数	30000
版次	2025年3月第1版
印次	2025年3月第1次印刷
标准书号	ISBN 978-7-5713-4945-5
定价	288.00元（精）

序

回想起来，我与曾建龙的相识，大概是在 17 年前，我 50 岁的时候，在马来西亚的海岛上。在那一趟旅程中，我们从设计、艺术、生活、天文、地理到人生，无所不谈。热情、幽默、率性，是我对这位年轻设计师的印象。大家都说，同频相应，或许这些性格就是我俩的契合点。

后来，我们一起去了巴厘岛，在异域谈论着对生活的感悟，也聊着对设计的理解。每次的相聚与对谈，我总会被他的创意、热情和真诚感染。记得曾有一次，我们在米兰家具展上偶遇，他兴奋地向我展示他的作品，希望我能给他一些建议，言语中流露出对设计的热爱和对未来的憧憬，让我也很受触动。直至今日，相识近 20 年，他对设计、生活依然充满热爱，不放过任何一个机会，不断学习，丰富自己。这么多年来，我看着他一步步变得成熟，从性格张扬的年轻设计师，到如今成为沉稳自信的跨界设计师，而这种蜕变也反映在他的作品中。我由衷地为他感到高兴，也为他的成就感到骄傲。

前些日子，他告诉我，他要出一本书，希望用自己的作品，记录这些年的成长和蜕变。我忽然间感慨万分，回想起相识的点滴，他做事有原则、有追求，这虽然让他一路走来经历了不少挑战和磨炼，但乐观的性格也令他可以从容面对。

作为设计师，我们都明白，设计不仅仅是美学，更是一种思考，是一种对生活的理解和对未来的展望。而曾建龙的作品，总是带着一种独特的气质。我想，这与他的性格、阅历脱不了关系，也是他对生活的感悟和对设计的追求。

我相信，这本书，正如同他一路走来见到的风景，是故事的记录，也是梦想的延续，也像他的人生一般，充满热情和对未来的期许。

致曾建龙，愿你不忘初心，永持热爱。

梁志天

2024 年 11 月

PREFACE

It is with great pleasure and deep respect that I write this preface to Gary Zeng's book, a collection that encapsulates not only his evolution as a designer but also the personal journey that has shaped his work and vision over the years. Our acquaintance spans nearly two decades, a friendship that began on a serene Malaysian island, where, as two individuals united by a shared passion for design, art, and life, we embarked on conversations that ranged from the cosmos to the intricacies of the human experience. The exchanges we had during that time left an indelible impression on me Gary's enthusiasm, humor, and straightforwardness were qualities that instantly resonated with me. The saying "birds of a feather flock together" could not have been more fitting in this context, as we discovered early on that our shared interests and outlook on life would forge a lasting bond.

Our journeys together continued to evolve, taking us from Malaysia to Bali, where our conversations shifted to an even deeper exploration of design and life's purpose. It was in these moments that I came to admire Gary's boundless creativity, passion, and sincerity. One particularly memorable encounter took place at the Milan Furniture Fair, where Gary, in his characteristic excitement, shared his latest work and sought my feedback. It was a moment that revealed not only his commitment to design but also his desire for constant self-improvement and his dedication to shaping the future of his craft. His openness and enthusiasm left a profound impact on me. Over the years, I have witnessed Gary's remarkable transformation—from a young designer full of vibrant energy to the composed and confident professional he is today, a designer whose work transcends boundaries. It has been a privilege to observe not only his professional success but also his growth, and I am genuinely proud of all that he has achieved.

Gary's decision to publish this book is a testament to his ongoing journey, a desire to document the milestones and experiences that have shaped his life and work. His growth has been neither linear nor without its challenges, yet his unwavering optimism and commitment to doing what is right have enabled him to overcome obstacles with grace and determination. In many ways, this book represents the culmination of those experiences—a reflection of both the man and the designer.

As designers, we recognize that our work is not merely about aesthetics. Design is a reflection of our understanding of the world, our vision for the future, and our approach to life itself. Gary's work embodies these principles in a way that is uniquely his own, each piece infused with his character, intellect, and passion. His creations are not simply objects, they are expressions of a deeply considered perspective on life.

This book is not just a record of Gary's career—it is a testament to the landscapes he has encountered, the challenges he has faced, and the dreams he continues to pursue. It is, above all, a reflection of his undying passion for design and his belief in future.It is an honor to contribute these words to the beginning of this book, and I have no doubt that Gary's journey, as captured here, will inspire all who encounter it.

To Gary Zeng—may your passion for life and design continue to guide you toward new horizons, and may this book be yet another step in your ongoing story of growth and discovery.

Steve Leung
November 2024

餐厅空间美学的探索与诠释

在过去的 10 年里，我们 GID 团队全身心投入餐厅项目设计的领域。从商业品牌模式的角度来说，消费者体验的差异化促使品牌空间形成区域的独特性。我们通过对空间的各种探索，不断挑战自我，追求创新，旨在为品牌和食客们打造出不仅满足味蕾，更能触动心灵的餐饮美学空间。

· 营造独特的用餐氛围

空间美学是我们设计的核心。我们深知，一个成功的餐厅不仅仅是提供美食的场所，更是一个能让人们沉浸其中，感受美的空间。在每一个项目中，我们精心考量建构的关系、光线的运用，通过巧妙的照明设计，营造出温馨、浪漫或热烈的氛围，以满足不同餐饮品类的需求。并且，注重空间尺度的比例和布局的合理性，确保食客在舒适的环境中享受美食。无论是宽敞的大厅还是私密的包间，都经过精心规划，以实现空间的最大化利用和最佳的人流导向。

· 餐饮品类与空间风格的融合

餐饮品类的多样性决定了空间风格的多元化。对于传统的中式餐厅，我们借鉴绘画、建筑等美学元素，将其融入空间，展现出优雅的当代东方韵味；而对于现代西式餐厅，我们则采用建构关系、色彩碰撞、多样的材质组合，营造出时尚、简约的氛围。无论是精致的法餐、南洋风味的泰国餐还是具有烟火气的火锅，我们都努力使空间风格与餐饮品类的特点相得益彰，让食客在踏入餐厅的瞬间，就能感受到空间的独特魅力。

· 文化主题的挖掘与呈现

文化主题是餐厅设计的灵魂所在。我们深入挖掘各地的地域文化特色，将其融入空间设计中。例如，在设计一家以闽南文化为主题的餐厅时，我们运用了闽南建筑、燕尾屋檐、南洋风情、六角红砖和剪纸等元素，打造出充满闽南文化气息的空间；而在一家以 7 幅名画为主题的餐厅里，我们则以名画的意韵来建构脉络，以画中的故事为灵感，编排舞蹈，让舞蹈和空间营造出东方美学般的诗意空间情绪，同时给用餐者提供精致的美学体验。通过文化主题的呈现，我们不仅为食客带来独特的用餐环境，更传承和弘扬了地域文化的魅力。

· 空间与时间的对话

我们在设计中会遇到一些有着历史时间轴的项目，在这样的项目中，我们尝试让空间与时间进行对话。通过保留老建筑的结构和元素，赋予其新的生命，展现历史与现代的交融。同时，运用现代材料和技术，创造出具有未来感的空间效果，让食客在同一空间中感受到时间的流转和变迁。这种对时间维度的探索，使餐厅不仅仅是当下的存在，更成为连接过去、现在和未来的纽带。

· 艺术性与摩登性的碰撞

艺术是提升空间品质的重要手段。我们与艺术家合作，在餐厅中展示绘画、雕塑等艺术作品，为空间增添浓厚的文化气息。同时，紧跟时尚潮流，将摩登的设计元素融入其中，使餐厅始终保持新鲜感和吸引力。在过去的十年里，我们不断探索艺术性与摩登性的完美结合，打造出既具有艺术价值又符合现代审美的主题餐厅。

本书精选近 4 年的优秀作品，是 GID 设计对餐厅空间美学的不断诠释和探索的见证。每一个项目都是一次挑战，也是一次成长。我们将继续秉持创新精神，不断挖掘空间的潜力，带来更多令人惊艳的餐饮空间设计。

曾建龙

2024 年 10 月于上海

EXPLORE AND INTERPRETE RESTAURANT SPACE AESTHETICS

In the past ten years, our team has been fully committed to the field of restaurant project design. Thinking from the perspective of business model: differentiating consumer's experience, forming regional uniqueness of brand space. Through various explorations of space, we keep to challenge ourselves and pursue innovation.We aim to create a catering aesthetic space for brands and diners that not only satisfies the taste buds but also touches the soul.

·Create a unique dining atmosphere

Spatial aesthetics is the core of our design. We deeply understand that a successful restaurant is not only a place to provide delicious food but also a space where people can immerse themselves and feel beauty. In each project, we carefully consider the constructed relationship and the use of light. Through ingenious lighting design, we create a warm, romantic or enthusiastic atmosphere to meet the needs of different catering categories. Moreover, we also pay attention to the proportionality of spatial scale and the rationality of layout to ensure that diners could enjoy delicious food in a comfortable environment. Whether spacious halls or private rooms, they are all under elaborate plans to achieve maximum utilization of space and the best traffic orientation.

·The fusion of catering categories and space style

The diversity of catering categories determines the diversification of space styles. In traditional Chinese restaurants, we draw on aesthetic elements such as painting and architecture,and integrated them into the space to show an elegant contemporary oriental charm; while in modern Western restaurants, we adopt constructed relationships, color collisions, and combinations of diverse materials to create a fashionable and simple atmosphere. Whether it is an exquisite French meal, Thai cuisine with Southeast flavor or hot pot which is full of earthly atmosphere, we all strive to make the space style and the characteristics of catering categories complement each other, so that the diners can feel the unique charm of the space since the moment they step into the restaurant.

·The excavation and presentation of cultural themes

Cultural theme is the soul of restaurant design. We deeply explore the regional cultural characteristics of various places and integrate them into the space design. For example, when designing a restaurant themed on Minnan culture of Fujian, we use elements such as Minnan architecture, swallowtail eaves, Asian style, hexagonal red bricks and paper-cuts to create a space full of Minnan cultural atmosphere; while in a restaurant themed on seven famous paintings, we use the charm of famous paintings to construct the context and draw inspiration from the stories in paintings to choreograph, so that dances and space can create an oriental aesthetic poetic field and emotional atmosphere, and at the same time provide diners with an exquisite aesthetic experience. Through the presentation of cultural themes, it not only brings a unique dining environment for diners but also inherits and promotes the charm of regional culture.

·The dialogue of space and time

We always encounter some projects with a long historical timeline. In such projects, we try to let the space have a dialogue with time. By preserving the structure and elements of old buildings and giving them new life, we show the blend of history and modernity. At the same time, we use modern materials and technologies to create a space effect with a sense of the future, allow diners to feel the change of time in the same space. This exploration of the time dimension makes the restaurant not only an existence in the present but also a link connecting the past, present and future.

·The collision of artistry and modernity

Art is an important means to improve the quality of space. We cooperate with artists to display artworks such as paintings and sculptures in restaurants, add a strong cultural atmosphere to the space. Meanwhile, we closely follow the fashion trends and integrate the modern design elements, to keep the restaurant always fresh and attractive. In the past ten years, we have been constantly exploring the perfect combination of artistry and modernity to create themed restaurants that not only have artistic value but also meet modern aesthetic needs.

This book includes the works in recent four years, is a witness to our continuous interpretation and exploration of restaurant space aesthetics. Each project is a challenge and also a growth. We will continue to uphold the spirit of innovation, constantly tap the potential of space, and bring more stunning catering space designs.

Gary Zeng

October 2024 in Shanghai

曾建龙（Gary Zeng）

GID 格瑞龙国际设计创始人、董事
融舍艺术家居品牌
意大利品牌 Roberto Cavalli 的合作设计师
自由艺术家

Founder and director of the GID Grignon International Design Co. Ltd.

Precious Home Art furnishing brand

The first Chinese designer to work with Roberto Cavalli, a brand in Italy

Freelance artist

曾建龙作为 GID 格瑞龙国际设计品牌的核心之一，除负责公司日常运营管理之外，更多的是做设计创意，把每一个项目或者产品都注入新的思想和设计表现，在团队中，他在定位和创造力方面发挥着核心作用。凭借在国外多年的学习和工作经验，曾建龙学会用多维视角去理解事物的差异性，喜欢将自己对生活的理解和感知用自己的方法融入项目，并最终呈现在作品中。

曾建龙的作品涉及建筑、产品、空间、微观视觉艺术、商业模式构造等领域。设计作品曾获得德国红点设计大奖、美国年度最佳设计奖金奖、意大利设计大奖赛金奖、英国餐厅和酒吧设计大奖、英国安德鲁 · 马丁国际室内设计大奖、韩国 K-DESIGN AWARD 等国际知名奖项。

曾建龙秉持对当代东方美学生活方式的研究精神，从空间设计到产品创作，不断探索、拓展设计领域。他善于将美学方式与商业模式相融合，并用多维视角去理解事物的差异化。设计源于生活，设计改变生活。

As one of the core figures of GID international design brand, Gary Zeng, in addition to the daily operation and management of the company, devotes more time to creative design. He infuses new ideas and design expressions into every case or product. Gary plays a core role in positioning and creativity within the team. With his years of study and work experiences abroad, Gary has learned to understand the differences of things from multiple perspectives. He likes to integrate his understanding and perception of life into projects in his own way and finally present them in his works.

Gary's works cover fields such as architecture, products, space, microscopic visual art, and business model construction. His design works have won international renowned awards such as the German Red Dot Award, the American Best of Year Awards Gold Medal, the Italian A'Design Award Gold Medal, the British Restaurant and Bar Design Awards, the British Andrew Martin International Interior Design Award, and the South Korean K-DESIGN AWARD.

Gary adheres to the research spirit of contemporary oriental aesthetic lifestyle, continuously explores and expands the field of design, from space design to product creation. He is good at integrating aesthetic methods with business models, and uses a multi-dimensional perspective to understand the differences of things. Design comes from life and design changes life.

鳳陽338

公司简介
ABOUT GID

GID 设计是一家集创意和商业模式形态于一体的设计公司，主营建筑设计、别墅景观设计、室内空间设计、产品设计。公司位于新加坡和中国上海，上海为运营总部，分公司所在城市有宁波、厦门、沈阳、无锡等。

GID 设计强调“生活艺术化，艺术生活化”的设计美学理念，并以多元化创意延续于不同领域的空间设计。GID 设计团队以商业模式创新和极富张力的设计表现而受到市场认可。以“高水平的商业模式创新品牌和空间设计服务价值”为核心，让每个项目呈现“有辨识度的商业空间形态、创造品牌用户体验价值及空间美学价值”。公司多年来服务于不同国际品牌和高品质客户群体，用创意和商业顶层模式设计理念来展示自身的服务价值。

GID 设计通过坚实的团队力量来实施完整的项目落地，同时联合众多专业顾问团队一起打造优质的项目。近些年 GID 设计致力于商业空间设计（特别是餐饮空间和生活品牌空间的打造），营造出精致多样的室内空间。GID 设计秉承设计不重复的美学理念，一直走在探索设计的美学之路上。GID 设计的服务范围包括家具产品设计、别墅空间设计、时尚精品酒店设计、餐饮空间设计及商业品牌空间设计。GID 设计以创新商业顶层模式的逻辑思维，打造室内商业空间的创新性。

GID Design is a design company integrating creativity and business model forms. It is a design and creative company integrating architecture, villa landscape, interior space and product design. The company is located in Singapore and Shanghai, China. Shanghai is the core of the operation company headquarters, and the branch cities include Ningbo, Xiamen, Shenyang, Wuxi, etc.

GID Design emphasizes the design aesthetic concept of "life artification and art lifeization", and continues in different field space designs with diversified creativity. The GID designer team is recognized by the market for its business model innovation and individualized tension design performance.Centering on "the high level of business model innovation brand and space design service value", it presents "distinctive commercial space forms, creates brand user experience value and space aesthetic value" for each project. The company has served different international brands and high-quality customer groups for many years. GID presents the service value of the GID design agency with creative and business top-level model design concepts.

We implement the complete project through the rigorous team strength. At the same time, we unite many professional consultant teams to create high-quality projects. In recent years, GID Design has been committed to commercial spaces (especially the creation of catering spaces and lifestyle brand spaces), creating exquisite and diverse interior spaces. We adhere to the aesthetic concept of no repetitive design and have been on the road of exploring the aesthetics of design.The service content scope of GID Design includes: furniture product design, villa space residence design, fashionable boutique hotels, catering spaces and commercial brand spaces. With the logical thinking of the innovative commercial top-level model, we control the innovation of indoor commercial spaces.

作品导览
CAPTIONS

曾建龙新作上海九储堂餐厅，从饮食的角度，运用色彩的变换诠释传统与现代的碰撞，实现束缚与释放的情绪上的平衡。

北京鲁采餐厅琨御府店与曾建龙携手，以宋代文化艺术为主线脉络，以古韵雅奢为氛围基调，共同打造出了一个沉浸式的具有宋代美学的餐饮空间。

曾建龙于广东中山打造的地标性高端餐饮“金汇酒楼”，以雅文化的当代性为核心，并从岭南建筑文化中提取设计形态语言，融入骑楼与镬耳墙等经典元素。

龙泰餐厅位于南京永隆 · 圣马可国际家居广场，是都市建筑群里突出的复古时髦风商业体。

上海老码头遗风鼎琪餐厅，出现在标志性的建筑里，历经数载依然被津津乐道。这里所承续的精神和美学，为老码头营造出活色生香的电影故事感。

PHOENIX 39 法餐厅位于合肥市行政中心区域，北面比邻当地最大的天鹅湖公园，西面是安徽广电中心，坐拥城市美丽的湖景。

余欢火锅餐厅坐落于南京老门东，是一座保护性历史建筑。青砖空斗墙，小青瓦坡屋面，南北向一进院落，两层木结构，呈现典型的南京门东传统院落民居特点。

曾建龙与西安它山石餐饮联手倾力打造的莱安五号餐厅，以东方美学为主轴，诠释盛唐文化的魅力风情。

在上海黄浦江湾北外滩的起点、苏州河外白渡桥畔、万国建筑群的延续地，有一处隐秘于繁华都市的桃花源——黛 · 外滩。

蟹仙画宴餐厅位于上海外滩五号，通过中国人耳熟能详的文化符号，食客们穿越时空来到千百年前，亲身感受、想象古代文人的生活方式。

曾建龙最新打造的餐饮空间“文杏馆 · 海”，以海派文化为出发点，解构海中波浪的形态，将充满韵律美感的纹样细节融入空间，营造极富视觉张力的用餐氛围。

福序餐厅以闽南菜系为特色，赋予上海滩滨江一道闽南美食文化风景线。福序餐厅的闽南文化符号性，作为一种凝固的艺术，赋予场域在地性的空间美学。

古运河承载着时光，深藏着城市的记忆，彰显着江南水乡的景色。无锡松果果餐厅是山脚下的一处隐藏美食之地，入院设有水景之境，竹林包裹着建筑的外立面，给人隐入山林的感受。

目录
CONTENTS

上海 · 外滩五号九储堂餐厅

THE NINE

A SELF-TALK IN A CENTURY-OLD BUILDING

在百年建筑里进行一场空间的自我对话

项目地点 : 上海　面积 : 600 m^2

Location：Shanghai　Area : 600 m^2

震旦

设计师曾建龙最新打造的九储堂餐厅位于上海外滩 5 号三楼。外滩 5 号毗邻黄浦江，从属于外滩建筑群，建筑特色鲜明，具有厚重的历史感及浓厚的文化气息。

一段十里洋场的场域印记，造就了上海外滩的别样面貌。从项目独特的地理优势入手，进而在海派文脉中找到了设计灵感，空间的色彩搭配，形成现代与复古恰到好处的结合，让新与旧的碰撞在空间中进行时空的对话，历史与当代交相辉映。

The Nine Restaurant recently created by Designer Gary Zeng is located on the third floor of No. 5 The Bund in Shanghai. The building is adjacent to the Huangpu River and is part of the Bund architectural complex. It has distinctive architectural features, a strong sense of history and a strong cultural attributes.

A special regional memory created a unique appearance of the Bund in Shanghai. Starting from the unique geographical advantage of the project, the designer found inspiration in the context of the Shanghai style. The color matching of the space creates a perfect combination of modern and retro, allowing the collision of the new and the old to have a dialogue of time and space, where history and contemporary complement each other.

平面图

FLOOR PLAN

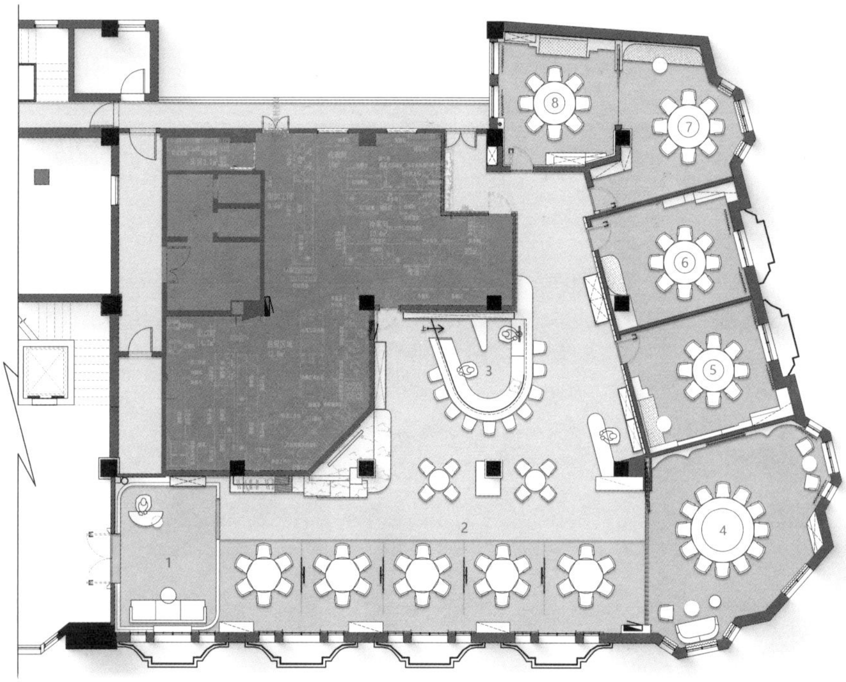

1.	前厅	LOBBY
2.	大厅	HALL
3.	酒吧	BAR
4.	包间 1	PRIVATE ROOM 1
5.	包间 2	PRIVATE ROOM 2
6.	包间 3	PRIVATE ROOM 3
7.	包间 4	PRIVATE ROOM 4
8.	包间 5	PRIVATE ROOM 5

绮丽浓烈

GORGEOUS AND INTENSE

MIRAE
震旦
花旗集团
citi

北京 · 鲁采餐厅琨御府店
LU STYLE

LU STYLE ENCOUNTERS
THE BEAUTY OF
THE SONG STYLE

鲁采遇见宋式美学，邂逅人文雅境

项目地点 : 北京　面积 : 2100 m^2

Location：Beijing　Area : 2100 m^2

鲁采

GID 对鲁采餐厅琨御府店的设计，不仅仅是对宋代美学的简单复制，更是对其精神内涵的深刻领悟与创新表达。它将宋代文化与现代生活完美融合，为宾客打造了一个沉浸式的餐饮空间，让人们在繁忙的都市生活中，找到一片诗意的栖居之地。

GID's design for Lu Style Kunyufu Store is not just a simple replication of Song Dynasty aesthetics, but a profound understanding and innovative expression of its spiritual connotation. It perfectly integrates Song Dynasty culture with modern life, creating an immersive dining space for guests, allowing them to find a poetic land in the busy urban life.

Modern
Plain

平面图
FLOOR PLAN

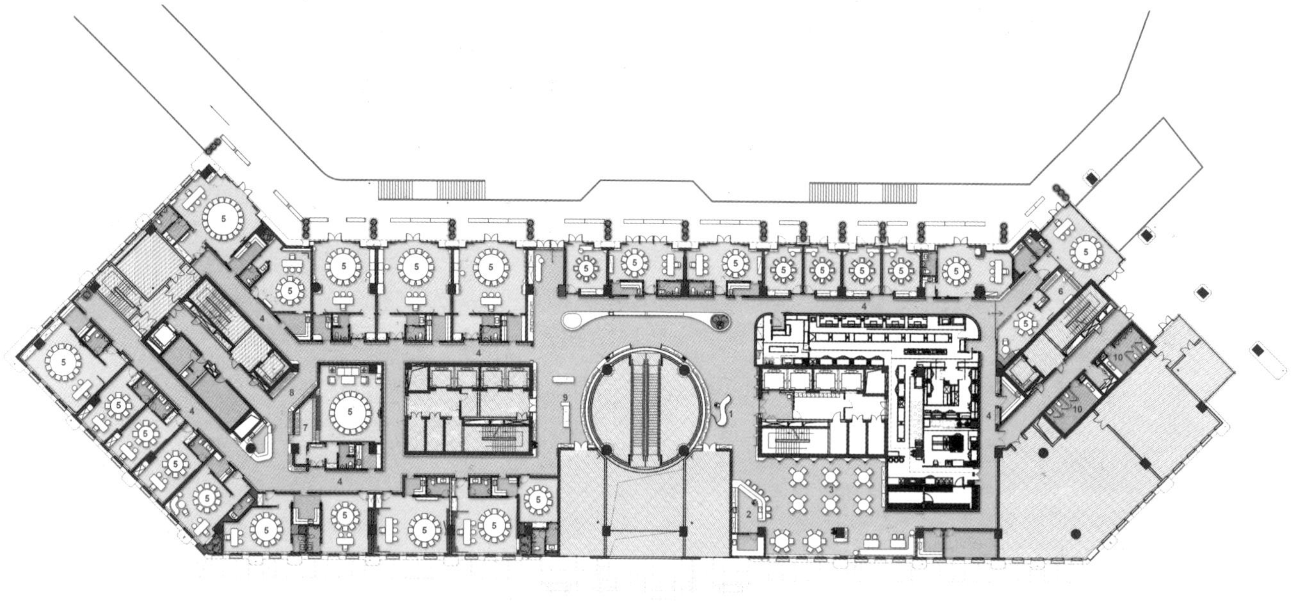

1. 前厅	LOBBY
2. 接待台	RECEPTION DESK
3. 海鲜池	SEAFOOD POOL
4. 收纳区	STORAGE AREA
5. 收银台、水吧	CASHIER & WATER BAR
6. 散座区	SCATTERED SEATS AREA
7. 舞蹈台	DANCE STAGE
8.VIP 包房 1	VIP PRIVATE ROOM 1
9.VIP 包房 2	VIP PRIVATE ROOM 2
10.VIP 包房 3	VIP PRIVATE ROOM 3
11.VIP 包房 4	VIP PRIVATE ROOM 4
12.VIP 包房 5	VIP PRIVATE ROOM 5
13.VIP 包房 6	VIP PRIVATE ROOM 6
14.VIP 包房 7	VIP PRIVATE ROOM 7
15.VIP 包房 8	VIP PRIVATE ROOM 8
16.VIP 包房 9	VIP PRIVATE ROOM 9
17.VIP 包房 10	VIP PRIVATE ROOM 10
18. 精神堡垒	SPIRITUAL FORTRESS
19. 茶区	TEA AREA
20. 周边售卖区	SURROUNDING SALES AREAS
21. 酒窖	WINE CELLAR
22. 休息区	REST AREA
23. 水景、等候区	WATERSCAPE & WAITING AREA

宋代，是中国文化艺术发展的巅峰时期，其美学理念以简约、含蓄、典雅著称。鲁采餐厅琨御府店的设计正是汲取了宋代文化艺术的精髓，并将其融入每一个细节之中。

The Song Dynasty was the peak period of Chinese culture and art, and its aesthetic concept is known for simplicity, implicitness and elegance. The design of the Lu Style Kunyufu Store draws on the essence of Song Dynasty culture and art and integrates it into each detail.

古韵雅奢

ELEGANT LUXURY WITH ANCIENT CHARM

对酒当歌

SINGING WHILE DRINKING

中山 · 金汇酒楼

JINHUI

MODERN DESIGN REVEALS THE CHARM OF TRADITIONAL LINGNAN ARCHITECTURE

当代设计诠释传统岭南建筑美学

项目地点：中山　面积：8000 m²

Location：Zhongshan　Area：8000 m²

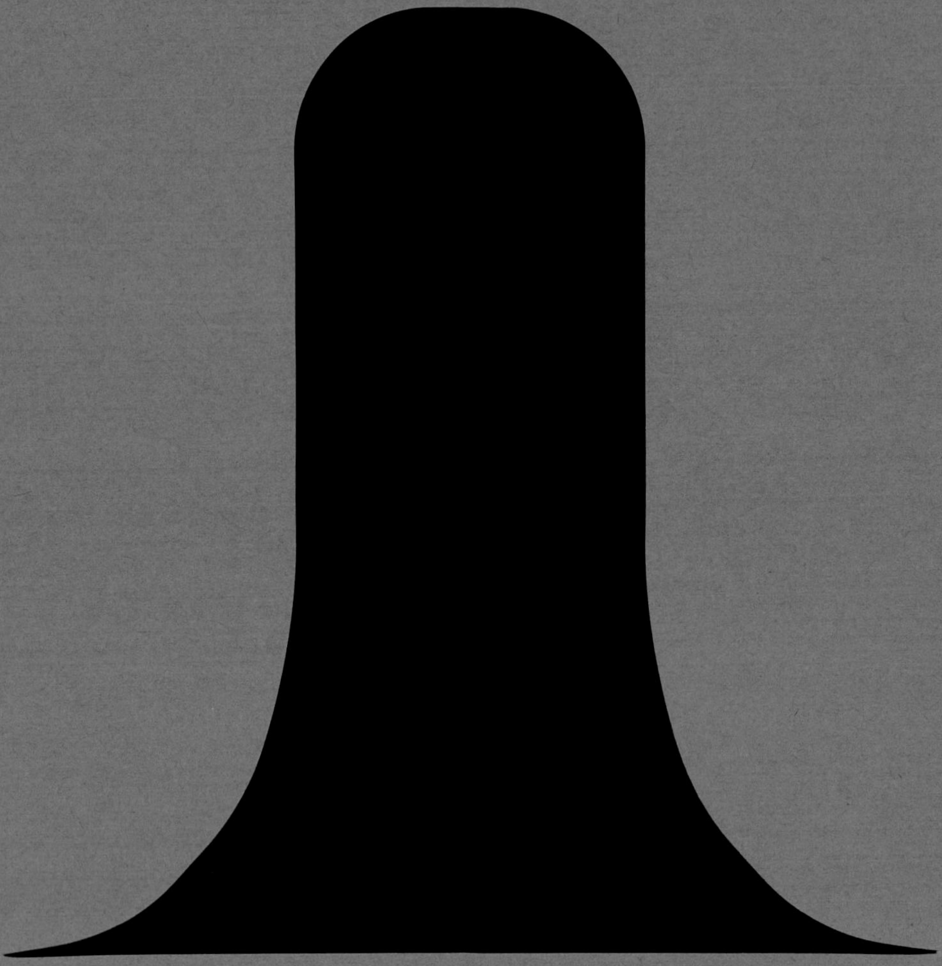

中山金汇酒楼总面积 8000m^2，配备大、中、小多个宴会厅及 30 个包间，满足婚庆喜事筹办及聚会宴请需求。

The total area of Jinhui Restaurant in Zhongshan is 8,000 square meters. The restaurant has multiple banquet halls of various sizes and 30 private rooms, meeting the needs of local people for wedding preparations and gatherings.

一层平面图

FIRST FLOOR PLAN

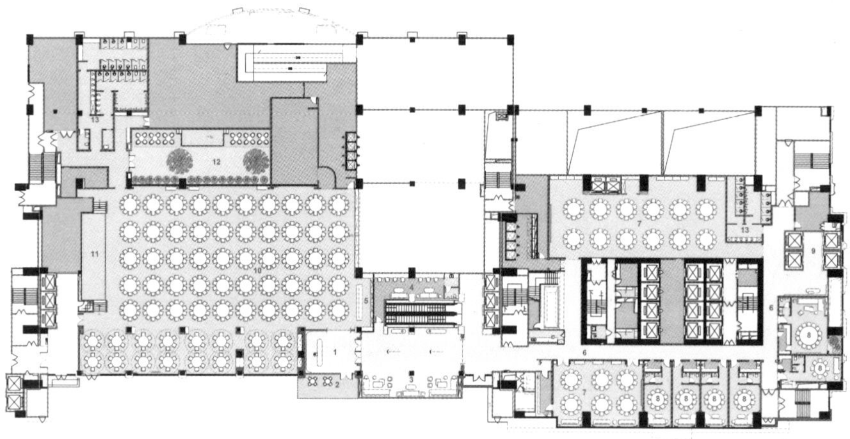

1. 接待台	RECEPTION DESK
2. 户外吸烟区	OUTDOOR SOMKING AREA
3. 等候区	WAITING AREA
4. 化妆间	DRESSING ROOM
5. 水吧	WATER BAR
6. 过道	AISLE
7. 宴会厅	BANQUET HALL
8. 包间	PRIVATE ROOM
9. 电梯厅	ELEVATOR HALL
10. 大宴会厅	GRAND BALLROOM
11. 舞台	STAGE
12. 户外景观	OUTDOOR LANDSCAPE
13. 公共卫生间	PUBLIC RESTROOM

二层平面图

SECOND FLOOR PLAN

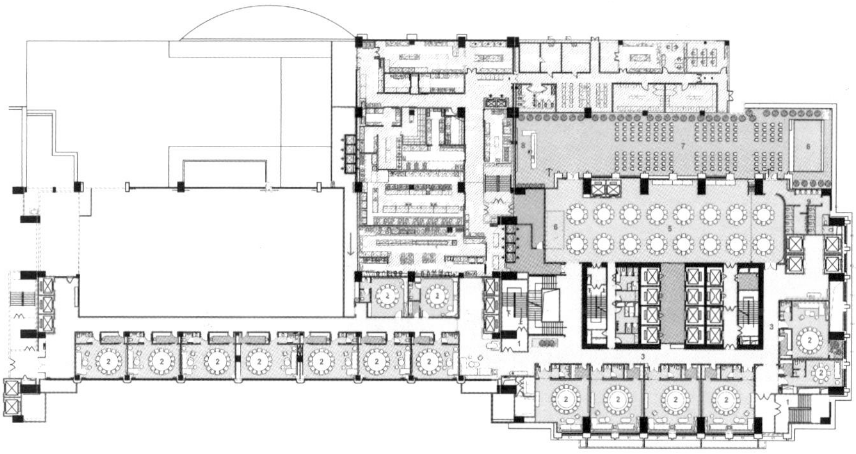

1. 楼梯间	STAIRWELL
2. 包间	PRIVATE ROOM
3. 过道	AISLE
4. 景观	LANDSCAPE
5. 宴会厅	BANQUET HALL
6. 舞台	STAGE
7. 户外花园	OUTDOOR GARDEN
8. 户外吧	OUTDOOR BAR
9. 公共卫生间	PUBLIC RESTROOM

设计以岭南建筑特点为基础，融合当代设计手法诠释空间美学，空间运用印度绿石材、金属、铝木结构的组合，打造出水墨画中苍松翠柏、山峦起伏、波光潋滟的幽远意境。

Based on the characteristics of Lingnan architecture, the design incorporates the contemporary design techniques to interpret the spatial aesthetics. The space used the combination of Indian green stone, metal corrugated panels and Aluminum wood structure to present the distant artistic conception of verdant pines and cypresses, undulating mountains and shimmering waves in ink wash paintings.

当代艺术

CONTEMPORARY ART

极简形态
MINIMALIST FORM

酒楼的每一层入口都设有独立门厅，彰显郑重的仪式感。

On each floor of the restaurant, there is an independent entrance hall at the entrance, showing a solemn sense of ceremony.

金汇酒楼整体空间设计以雅奢时尚、岭南特色主题、风格多元化为核心要素，延续中国古代文人雅士的风雅意趣，同时引入当代的时尚审美主张，在不同的时空背景下解读传统文化的魅力。

In the overall space design of Jinhui Restaurant, elegance and fashion, Lingnan characteristic themes and diversified styles are the core elements. The design continues the grace and interests of traditional Chinese literati and scholars, while introducing contemporary fashion aesthetic, interpreting the charm of traditional culture in different time and space backgrounds.

南京 · 龙泰餐厅
DRAGON SAWASDEE

BUILDING A MODERN THAI RESTO IN THE NEW BUSINESS COMMUNITY
在新商业群体中打造泰国餐厅的当代语境

项目地点：南京　面积：280 m^2

Location：Nanjing　Area : 280 m^2

普吉

長灘

龙泰餐厅（Dragon Sawasdee）位于南京永隆·圣马可国际家居广场，是都市建筑群里突出的复古时髦商业体。

Dragon Sawasdee Restaurant is located in Nanjing Yonglong San Marco International Home Furnishing Plaza, a prominent retro and fashionable commercial complex in the urban building complex.

平面图

FLOOR PLAN

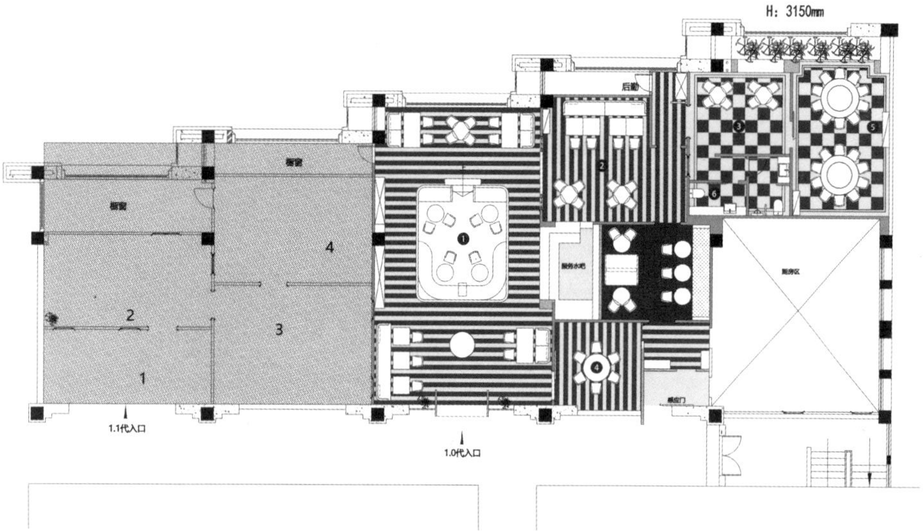

1. 散座区　SCATTERED SEATS AREA
2. 散座区　SCATTERED SEATS AREA
3. 散座区　SCATTERED SEATS AREA
4. 散座区　SCATTERED SEATS AREA
5. 包间 1　PRIVATE ROOM 1
6. 包间 2　PRIVATE ROOM 2
7. 卫生间　TOILET

设计师从泰国的风土人情中汲取灵感，将宫殿、花海、吊床、英伦形态等元素，以魔法师的手法巧妙而自然地融合到餐厅空间里，营造出别具一格的泰式意境。

The designer inspired by Thai culture and art, skillfully and seamlessly integrated elements such as palaces, flower fields, hammocks, and British styling into the restaurant space. This thoughtful incorporation makes a distinctly Thai artistic environment.

龙泰餐厅的整体设计，主要以黑白为空间底色，同时将典型的泰式风格色彩淋漓尽致地表现出来。浓烈的红、黄、绿、蓝，与简约的黑白色相碰撞，形成强烈的视觉冲击力和丰富的空间层次。

The design of Dragon Sawasdee Restaurant utilizes black and white as the primary background colors of the space. This color scheme can help to enhance vibrant Thai-style hues, including bold red, yellow, green, and blue. The dynamic interplay between these vivid colors and the understated black-and-white backdrop creates a striking visual impact and adds the space hierarchy.

鲜活之美

THE BEAUTY OF FRESHNESS

上海·鼎琪餐厅外滩店
JADE VESSEL

AN ORIENTAL FEAST OF SHANGHAI SHIKUMEN CULTURE
上海石库门文化的海派美食盛宴

项目地点 : 上海　面积 : 1100 m²

Location：Shanghai　Area : 1100 m²

JADE VESSEL

鼎琪餐厅位于上海外滩老码头 3 号库，在这个空间内，利用设计来营造具有海派特色的故事场景。从餐厅入口开始，超高动态数字化屏幕上的江流山川景象令人沉浸其中，忘却他物。

JADE VESSEL Restaurant is located at the Cool Docks, Bank 3. In this space, the designer used design to create a scene with the story of ancient and modern Shanghai style. From the restaurant's entrance, the depiction of rivers on an ultra high definition digital screen immerses viewers visually and makes them forget everything else.

平面图

FLOOR PLAN

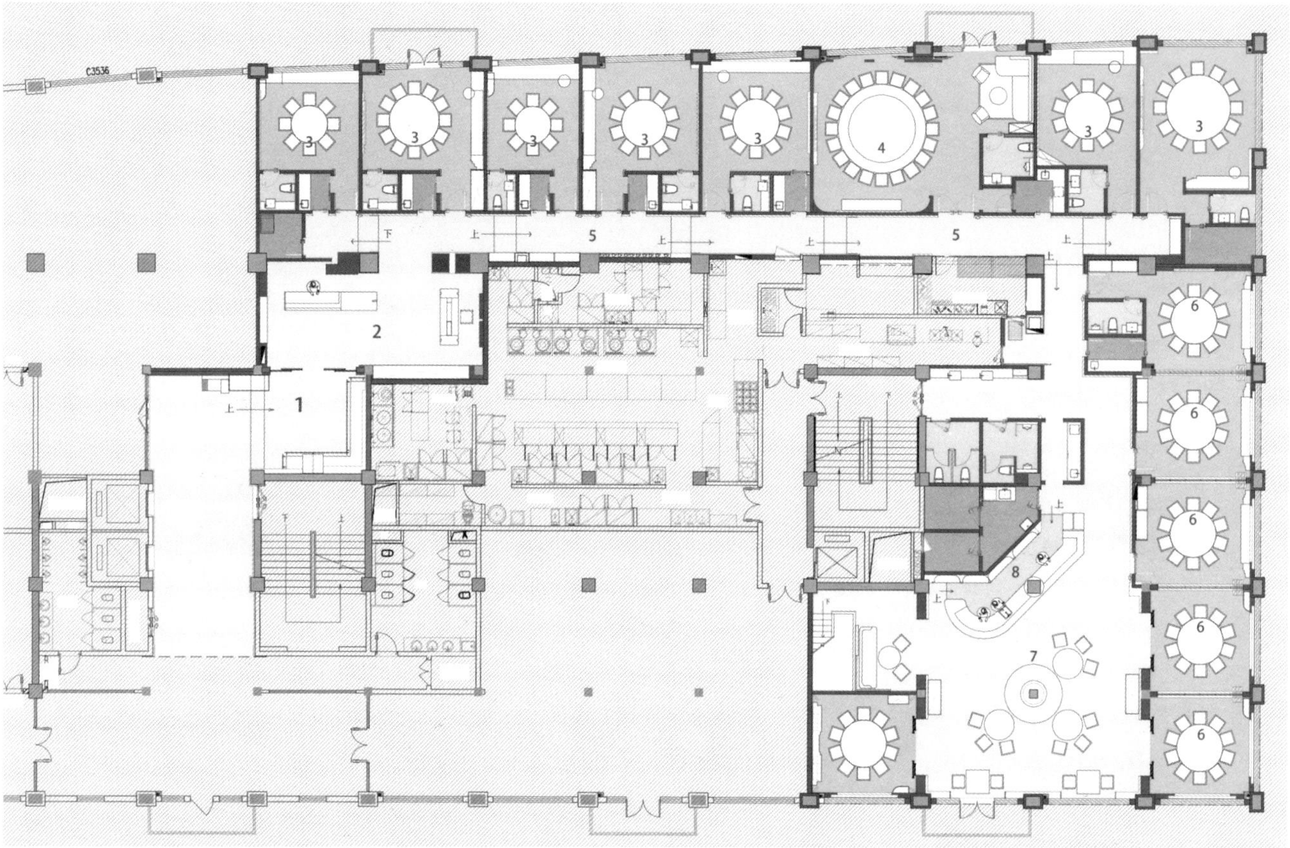

1. 前厅区　LOBBY
2. 接待台　RECEPTION DESK
3. 包间　PRIVATE ROOM
4. VIP 包间　VIP PRIVATE ROOM
5. 过道　CORRIDOR
6. 五连包间　FIVE NO- PARTITION PRIVATE ROOMS
7. 大厅散座区　SCATTERED SEATS AREA
8. 水吧　WATER BAR
9. 卫生间　TOILET

运用场地的时间脉络，演绎出古典与现代交织、张力与细腻重叠、传统与时尚融合的空间设计，赋予空间无尽的魅力。

By leveraging the temporal relationship of the space, the overall design interpreted the intersection of classical and modern elements, along with cultural heritage and contemporary influences, endowing this space with unparalleled charm and imagination.

气韵上，设计师融合工业的时尚元素，并保留空间原有的痕迹，通过颜色的对比、肌理的呈现，以及墙面构造的设计，将“光和空间”进行多维度的演绎，呈现别样的张力维度与感官体验。

In terms of charm, the designer combines the industrial fashion elements and retains the original traces of the space. Through the contrast of colors, the presentation of texture, and the design of the wall structure, the "light and space" is interpreted in different dimensions, presents a special tension and sensory experience.

百转千回
FULL OF TWISTS AND TURNS

鼎琪烟火

JADE VESSEL'S FIREWORKS

合肥 · PHOENIX 39 法餐厅

PHOENIX 39

FRENCH CUISINE MEETS MODERN DREAM OF BUTTERFLIES AND FLOWERS

法餐遇见蝶恋花的摩登梦境

项目地点 : 合肥　面积 : 100 m^2

Location: Hefei　Area : 100 m^2

PHOENIX39

从将餐厅主题定为“法式浪漫”开始，空间的造型艺术势必会拥有无可比拟的表达方式。

Given that the restaurant is characterized by French Romance, the art of the space must embody an unparalleled expression.

时尚、潮流、前卫、美丽、创新，这些汇集在当代年轻人思想中的关键词，拓展了设计师对空间情感的阐述。

These keywords such as stylish, trendy, avant garde, aesthetic, and creative highly valued by contemporary youth guide the designer's approach to interpreting spatial emotions.

设计师曾建龙“庄周梦蝶”，在空间奇境与现实生活之间，用法式主义彻底解放浪漫内核，看似脱离现实的艺术方式，让人迷失奇幻国度，打造一座城市中最具标志性的餐厅之一。

Designer Gary Zeng completely liberates the romantic core between the wonderland of space and real life. The seemingly unrealistic art style makes people lose their way in the fantasy while creating one of the most iconic restaurants in the city.

平面图

FLOOR PLAN

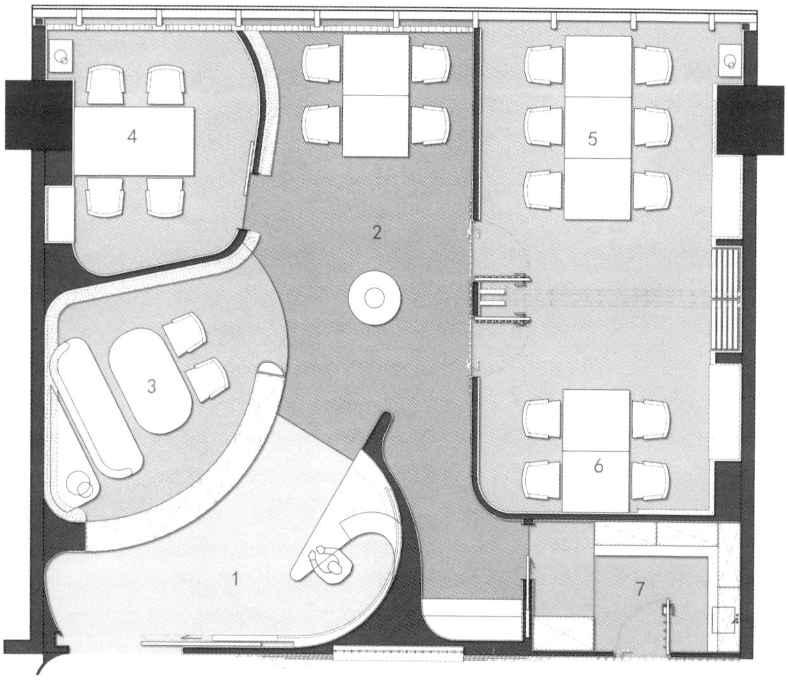

1. 入口　ENTRANCE
2. 大厅　HALL
3. 卡座　BOOTH
4. 四座包房　FOUR- SEAT PRIVATE ROOM
5. 包间　PRIVATE ROOM
6. 包间　PRIVATE ROOM
7. 餐具室　PANTRY

当人们共享一段旅程，从相聚到分别，彼此心里会留下什么样的印记呢？或许是勃艮第红酒炖牛肉散发的醇香、法式烤布蕾的甜糯，抑或是鱼子酱的独特口感。

When people share a journey, from gathering to parting, what kind of mark is left in each other's hearts? Perhaps it will be the mellow aroma of Burgundy beef stew, the sweet and glutinous taste of French creme brulee, or the uniqueness of caviar.

蝶恋

LOVE FOR BUTTERFLY

PHOENIX39

花海

FLOWER SEA

印象
IMPRESSION

南京 · 余欢火锅餐厅
YUHUAN

TELLING A JIANGNAN DREAM IN A CENTURY-OLD HUI-STYLE BUILDING
在百年徽派建筑里讲述一场江南梦境

项目地点：南京　面积 : 650 m^2

Location：Nanjing　Area : 650 m^2

YUHUAN

在南京老门东百年徽派建筑里，设计师曾建龙演绎了一场设计梦境。设想在余欢火锅餐厅的中式建筑空间中，一位身穿旗袍穿越时空而来的女子行走于二楼廊道的同时，一位现代男子步入了厅堂，来自不同时空的两人邂逅了。

In the century-old Hui-style building, designer Gary Zeng created a design dream in Yuhuan Hot Pot Restaurant. In the field space of this Chinese architecture, a lady in traditional cheongsam walking through the corridor on the second floor and a modern gentleman entering the lobby, representing a convergence of different times and spaces.

一层平面图
FIRST FLOOR PLAN

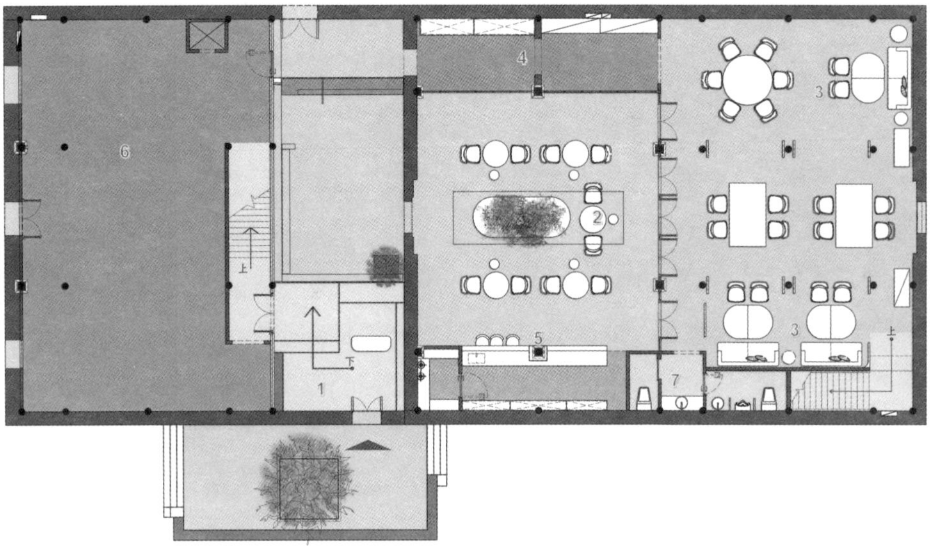

1. 前厅区 LOBBY
2. 外摆区 OUTDOOR HALL AREA
3. 散座区 SCATTERED SEATS AREA
4. 备餐区 PANTRY
5. 吧台区 BAR AREA
6. 厨房 KITCHEN
7. 卫生间 TOILET

二层平面图
SECOND FLOOR PLAN

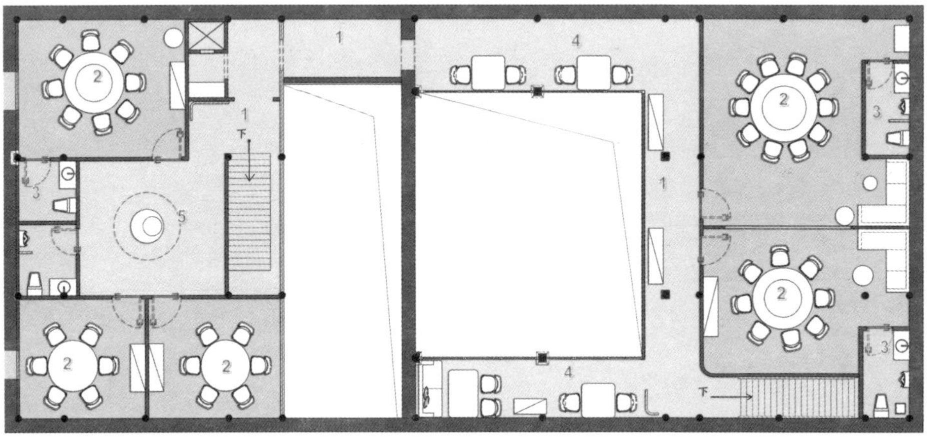

1. 廊道区 CORRIDOR
2. 包间 PRIVATE ROOM
3. 卫生间 TOILET
4. 散座区 SCATTERED SEATS AREA
5. 中厅区 ATRIUM

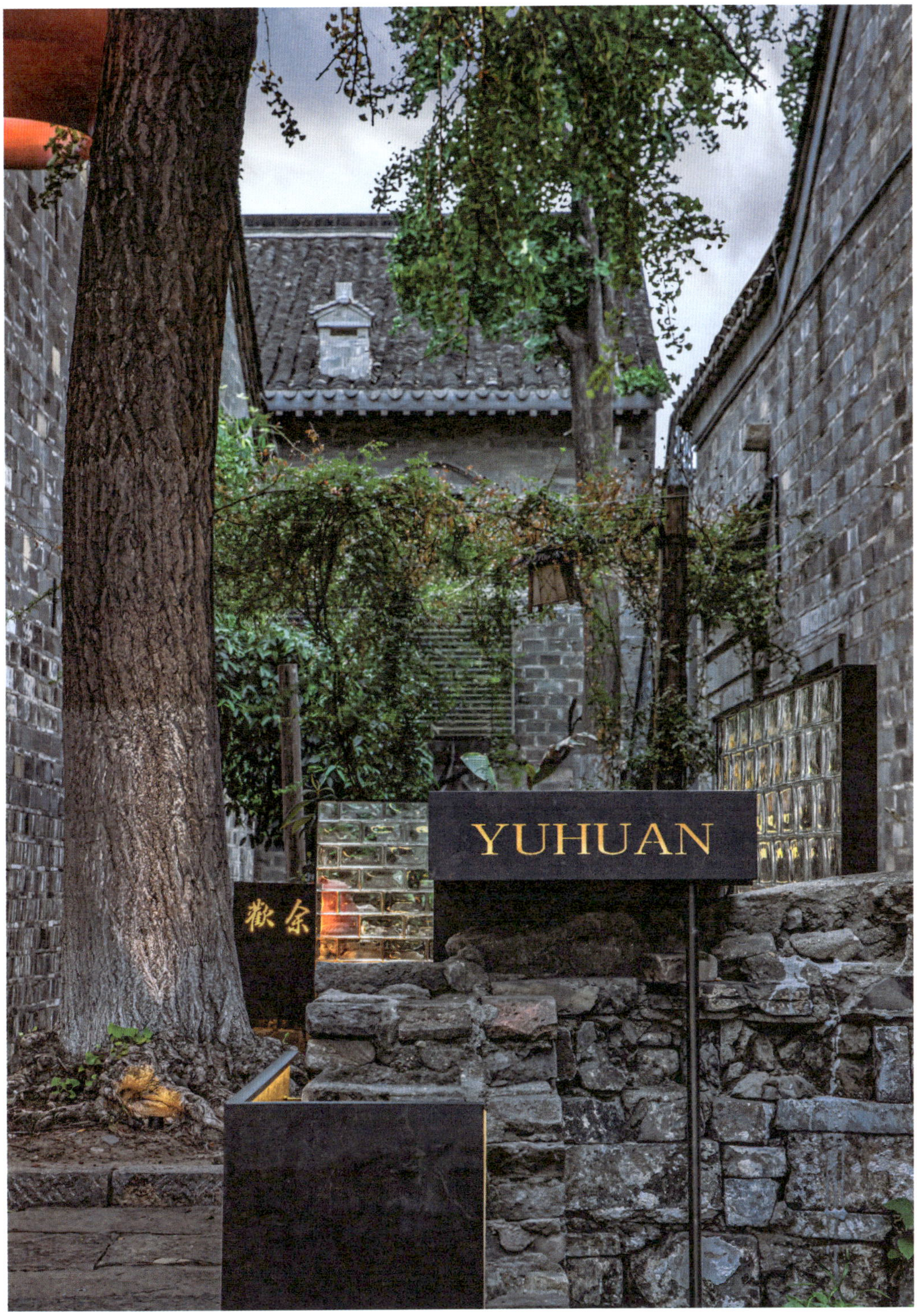

在这座位于金陵城老门东的百年建筑里，曾建龙以修旧如旧的手法，置身于时间长河中与空间对话，打造古今转场的“新盒子”。

In this historic building, Gary Zeng has preserved the building's original traditional style through artistic vision and techniques, allowing the past and present to coexist harmoniously and thereby creating a novel spatial experience.

推开镂空花格门，跨过一楼大厅的木门槛，进入“水与火”元素碰撞的空间。在这里，水与火相容，和谐共存。天花板上凹凸的水波纹，通过手工艺术漆涂刷，塑造出凝固的波浪纹理，呼应余欢海鲜餐厅的主题。

Push the hollow flower lattice door, cross the wooden threshold of the first floor hall, and enter the space where "water and fire" elements collide.Here,water and fire are compatible and coexist harmoniously. The concave and undulating water ripples of the ceiling are painted with the art lacquer by hand to create a solidified wave texture and reflect the theme of Yuhuan Seafood Restaurant.

百味余欢

ALL FLAVORS LINGERING JOYS

宴乐雅集

BANQUET AND ELEGANT GATHERING

西安 · 莱安五号餐厅
LAI'AN NO.5

THE CULTURE
OF TANG DYNASTY
IN LAI'AN NO.5

盛唐文化的当代诠释，讲述莱安五号的美学语境

项目地点 : 西安　面积 : 2450 m^2

Location：Xian　Area : 2450 m^2

莱安五号
LAI'AN NO.5

莱安五号餐厅位于西安莱安大厦五楼，其整体空间按照一步一景的美学定位打造，让人想起了杜甫的诗句“穿花蛱蝶深深见，点水蜻蜓款款飞”。超过 2000 m^2 的空间中共设有雅席 32 位、包间 14 个、户外庭院席位 20 位、菜道茶席 7 席。

Lai'an No.5 is located on the fifth floor of Lai'an Building, Xi'an. The overall space of Laian No.5 is diversified, with a total of 32 casual seats, 14 private rooms, 20 outdoor seats and 7 tea tasting seats, covering an area of over 2,000 square metres.

平面图

FLOOR PLAN

1. 电梯厅	ELEVATOR HALL
2. 前厅区	LOBBY
3. 服务前台	SERVICE FRONT DESK
4. 大厅散座区	SCATTERED SEATS AREA
5. 包间	PRIVATE ROOM
6.VIP 包间	VIP PRIVATE ROOM
7. 特色包间	SPECIAL PRIVATE ROOM
8. 酒窖	WINE CELLAR
9. 休闲区	LEISURE AREA
10. 茶艺展示	TEA ART EXHIBITION
11. 茶室	TEA ROOM
12. 露台	TERRACE
13. 户外吧	OUTDOOR BAR

莱安五号餐厅可谓西安城内极其稀有的一方净土，是一个让人流连忘返的都会桃源，也是一个“四时佳兴与人同”的觅食秘境。每一位来宾可以在此“乐飨”“品珍”“慢食”“莱茗”“雅聚”“重欢”，欢享华贵雅致的“长安新里风尚馔宴”。

Laian No.5 is a delightful space immersed in nature in Xi'an, offering a retreat where one can linger and connect with friends throughout any season. Irrespective of the passage of time, every guest can find solace here by drinking tea, dining, meeting friends, engaging in conversation, and savoring leisurely moments in this elegant restaurant.

吧台顶部的灯饰采用两个动态的龙形装置使空间更显灵动。大厅顶部通过不同色彩的镜面金属结构，带给人虚幻的感官体验，增强了空间的活力与张力。

The lighting on the top of the bar uses two dynamic dragon shaped devices to make the space more flexible. The illusory sensory experience generated by the mirrored metal structure at the top of the hall, with its varied colors, creates an effect that transports individuals to a different space and time. Concurrently, this design element enhances the vitality and tension of the space.

莱安五号的包间名称取自中国古代四时时序，意境深远。那时的诗人胸怀、画者情思与皇朝气韵，穿越千年来到了面前。

The name of the private room in laian No.5 is taken from the December of the four seasons in ancient China, which has a long history of cultural artistic conception. At that time, the poet's mind, the painter's feelings and the spirit of the emperor are progressing through thousands of years.

遐想莱安
REVERIE LAIAN

上海·黛·外滩餐厅
DIANE FINE CUISINE

A HIDDEN PARADISE IN THE CITY, DIANE FINE CUISINE
江南都市里的桃花源，黛·外滩餐厅

项目地点 : 上海　面积 : 600 m^2

Location：Shanghai　Area : 600 m^2

黛
外滩
DIANE FINE CUISINE
精制粤菜江南烩宴

在上海黄浦江湾北外滩的起点、苏州河外白渡桥畔、万国建筑群的延续地，有一处隐秘于繁华都市的桃花源——黛·外滩。

At the starting point of the North Bund of the Huangpu River Bay in Shanghai, by the Garden Bridge of Shanghai and the Suzhou River, the continuation of the world Expo of architectures, there is a hidden paradise in the bustling city: Diane Fine Cuisine the Bund.

江南的诗意栖居与精致美学、当代艺术与中国美食、文人雅集与艺术志趣，在这里深度融合。区别于传统世俗餐厅的定义，曾建龙将黛·外滩打造为都市中的“江南秘境与外滩桃花源”，充满诗意。

The poetic ambiance and refined aesthetics of Jiangnan, contemporary art and Chinese cuisine, literati elegance and artistic allure all integrated deeply. Deviating from the conventional notion of a typical restaurant, Gary Zeng made Diane Fine Cuisine to be a Jiangnan Mystery Land and a Peach Blossom Garden on the Bund in the city, embodying a captivating poetic vision.

平面图
FLOOR PLAN

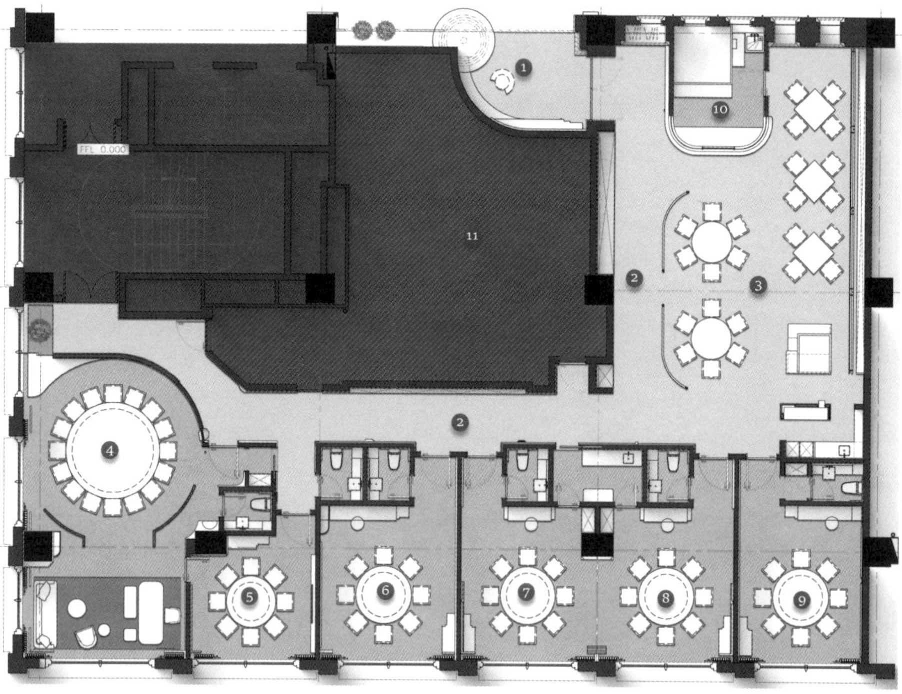

1. 前厅　LOBBY
2. 走廊　CORRIDOR
3. 散座区　SCATTERED SEATS AREA
4. 包间 1　PRIVATE ROOM 1
5. 包间 2　PRIVATE ROOM 2
6. 包间 3　PRIVATE ROOM 3
7. 包间 4　PRIVATE ROOM 4
8. 包间 5　PRIVATE ROOM 5
9. 包间 6　PRIVATE ROOM 6
10. 烤鸭房　ROAST DUCK ROOM
11. 厨房　KITCHEN

曾建龙擅长取景，本次设计对话商业与风尚，进行现代艺术与文化场域的重塑，利用多媒体艺术的展现、烤鸭房舞台的感观互动，聚焦一场颇具美食情绪价值的体验。空间风格以当代融合和海派调性为主，古典与摩登交织，张力与细腻重叠，国风与国际感融合，赋予空间无尽的魅力与想象力。

Gary Zeng stands out as a scene, talking with business and fashion, remodeling modern art and cultural fields, displaying multimedia art, focusing on an experience with culinary and emotional value. The style of the space is mainly based on contemporary integration and Shanghai style. Classical and modern interweave, tension and exquisite overlap, inheritance and international sense of fusion, giving the space endless charm and imagination.

翠微黛影

GREEN HILLS AND SHADOWY HUES

上海·外滩五号蟹仙画宴餐厅
CRAB PAINTING

A TRAVEL OF
PAINTING CUISINE AND FOOD
IN SEVEN RENOWNED PAINTINGS
穿梭在七幅传世名画中的画菜美食之旅

项目地点：上海　面积：600 m^2

Location: Shanghai　Area : 600 m^2

蟹仙画宴
沪上画中宴
世境胜文焉
馨食须臾间
美艺久长仙
蟹仙画宴
蟹仙画宴

轻灵瑞鹤

ELEGANT AND AUSPICIOUS CRANE

沪上画中宴，世境胜文焉。馨食须臾间，美艺久长仙。

Experience Infinite Beauty in Banquets in Shanghai Paintings.

蟹仙画宴
馨食须臾间
美艺久长仙
蟹仙画宴

蟹仙画宴位于上海外滩五号二楼，主打文化美食菜系。餐厅以七幅国宝级传世名画为灵感，融入七道画菜、七支画舞，将蟹宴的美食料理与中国传统文化相结合，呈现全新的“名画美学菜”。

The Crab & Paintings restaurant is located on the second floor of No. 5 The Bund in Shanghai, specializing in cultural and culinary cuisine. The restaurant draws inspiration from seven national treasures of renowned paintings, which serve as the essence of its concept. It integrates seven artistic dishes and seven performances into the dining experience, merging fine food with classical Chinese culture to create a novel Aesthetic Cuisine of Famous Paintings.

千里江山图

平面图

FLOOR PLAN

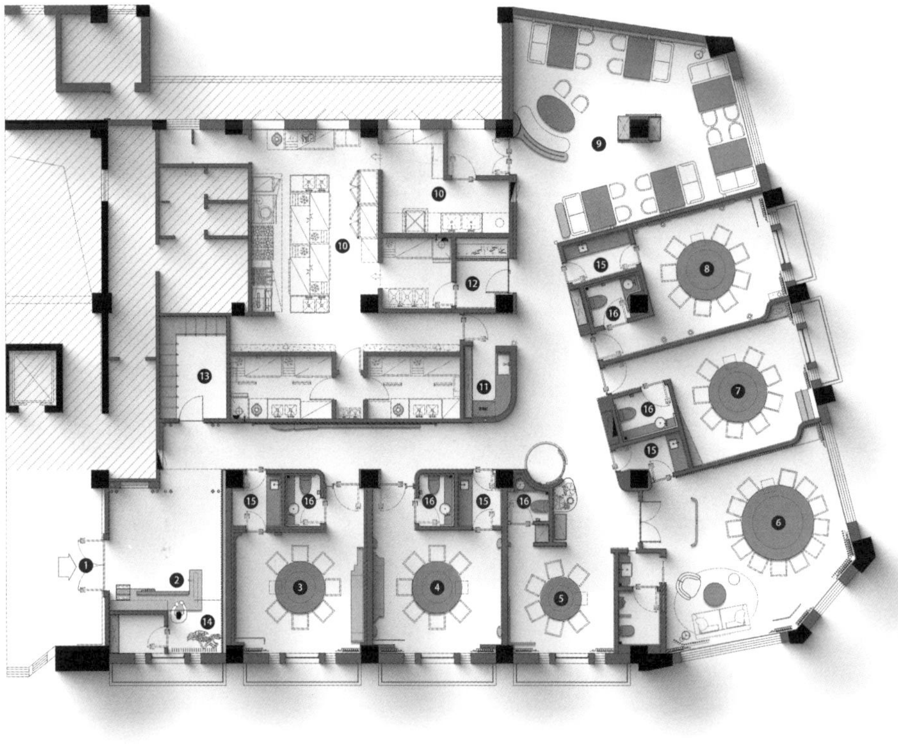

1. 入口	ENTRANCE
2. 接待台	RECEPTION DESK
3. 包间 1	PRIVATE ROOM 1
4. 包间 2	PRIVATE ROOM 2
5. 包间 3	PRIVATE ROOM 3
6. 包间 4	PRIVATE ROOM 4
7. 包间 5	PRIVATE ROOM 5
8. 包间 6	PRIVATE ROOM 6
9. 散座区	SCATTERED SEATS AREA
10. 明档	COOKING STATION
11. 茶区	TEA AREA
12. 海鲜室	SEAFOOD ROOM
13. 酒窖	WINE CELLAR
14. 储藏室	STORAGE ROOM
15. 备餐区	PANTRY
16. 卫生间	TOILET

通过中国人耳熟能详的文化符号，食客们穿越时空来到千百年前，亲身感受、想象古代文人的生活方式。

By harnessing the essence of traditional culture, it evokes an emotional resonance that transports viewers through time and space, allowing them to genuinely experience the artistic atmosphere of a thousand years ago.

复归风雅

RETURN TO ELEGANCE

MIRAE ASSET
震旦

千里江山

A THOUSAND MILES OF RIVERS AND MOUNTAINS

广州 · 文杏馆 · 海餐厅
MAISON GINKGO

A FEAST OF SHANGHAI STYLE CULTURAL CUISINE

文杏馆 · 海餐厅，演绎一场海派文化美食盛宴

项目地点 : 广州　面积 : 860 m^2

Location：Guangzhou　Area : 860 m^2

设计师曾建龙在广州海珠广场打造的餐饮空间“文杏馆 · 海”，以海派文化为出发点，解构海中波浪的形态，将充满韵律美感的纹样融入空间设计细节，营造极富视觉张力的用餐氛围。

The latest dining space Maison Ginkgo, created by designer Gary Zeng at Haizhu Square in Guangzhou, takes Shanghai culture as its starting point, deconstructs the shape of waves in the sea, and integrates rhythmic and beautiful pattern details into the space to create a dining atmosphere with great visual tension.

一层平面图
FIRST FLOOR PLAN

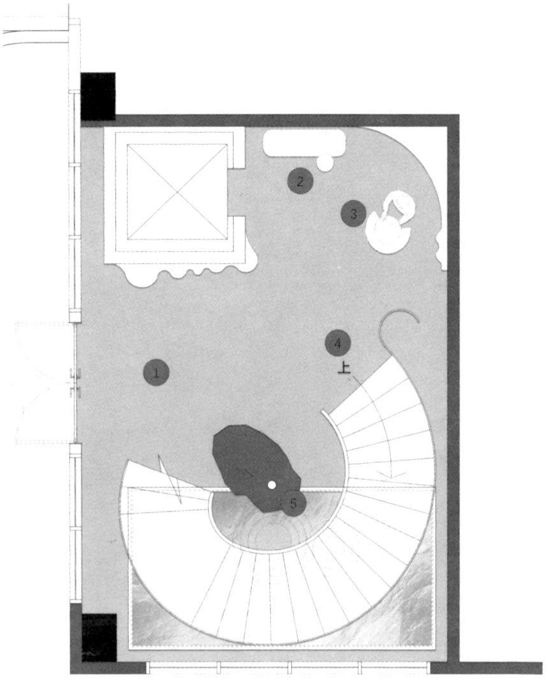

1. 入口 ENTRANCE
2. 电梯厅 ELEVATOR HALL
3. 接待台 RECEPTION DESK
4. 楼梯厅 STAIR HALL
5. 景观 LANDSCAPE

二层平面图
SECOND FLOOR PLAN

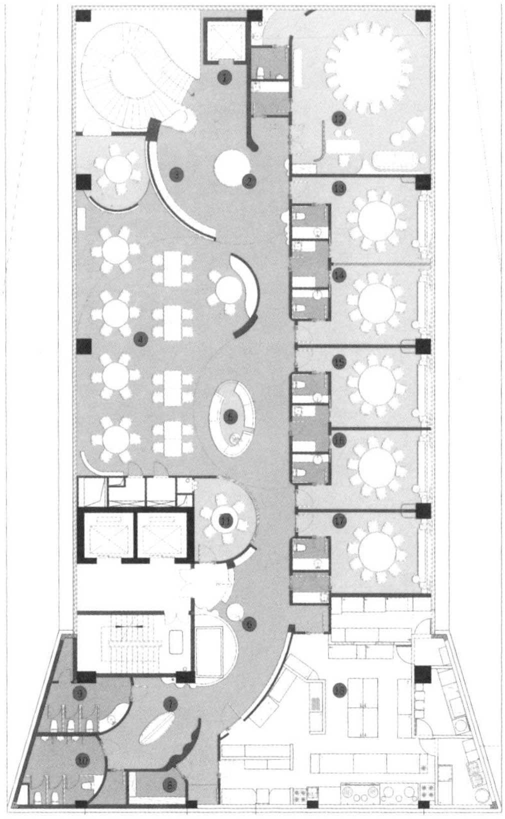

1. 入口 ENTRANCE
2. 水景 WATERSCAPE
3. 酒柜 GRADEVIN
4. 散座区 SCATTERED SEATS AREA
5. 服务岛台 SERVICE ISLAND
6. 艺术装置 ART INSTALLATION
7. 洗手台 WASH BASIN
8. 更衣室 LOCKER ROOM
9. 女卫 FEMALE TOILET
10. 男卫 MALE TOILET
11. 包房 7 PRIVATE ROOM 7
12. 包房 1 PRIVATE ROOM 1
13. 包房 2 PRIVATE ROOM 2
14. 包房 3 PRIVATE ROOM 3
15. 包房 4 PRIVATE ROOM 4
16. 包房 5 PRIVATE ROOM 5
17. 包房 6 PRIVATE ROOM 6
18. 厨房 KITCHEN

在海派特色文化餐饮空间“文杏馆 · 海”的整个设计过程中，曾建龙以海派文化底蕴为设计主线，深入挖掘和展现海派文化的独特魅力，希望通过这个项目向大众呈现一种独具特色的海派东方设计风格。

In designing the Shanghai style cultural dining space Maison Ginkgo, Gary Zeng focused on Shanghai's cultural heritage, delving deeply into its unique charm. The goal is to showcase a distinctive Shanghai style Oriental design and present a unique cultural experience to the public through this project.

MAISON GI

KGO

邂逅文杏

MEET THE GINKGO

见山见海

SEE THE MOUNTAINS AND THE SEA

上海 · 福序餐厅
ORDER BLESSING

MINNAN CULTURE INTERPRETS THE FEAST OF FUJIANESE CUISINE
闽南文化美学诠释闽菜盛宴

项目地点 : 上海　面积 : 600 m²

Location：Shanghai　Area : 600 m²

福
序
ORDER BLESSING

福序餐厅以闽南菜系为特色，赋予上海滩滨江一道闽南美食文化风景线。福序的闽南文化符号属性，作为一种凝固的艺术，赋予场域在地性的空间美学。

Order Blessing features Minnan cuisine and endows the riverside of Shanghai Bund with a Minnan cuisine cultural landscape. The symbol of Minnan culture in Order Blessing, as a vessel for history, confers localized spatial aesthetics to the field.

在地方菜系蓬勃发展的当下，既有山路菜与水路菜的山海向荣，又有官府菜、民间家常菜的上敬下和。这一次，古今东西兼容并蓄的闽菜，在异乡上海讲了一则开山出海的福建故事。

In the flourishing development of local cuisine today, a diverse array of dishes has emerged, blending sophisticated high end offerings with everyday fare. Fujianese cuisine, known for its integration of various ethnic specialties with traditional Chinese culinary traditions, exemplifies this trend. Shanghai Order Blessing Restaurant showcases the evolution of Fujianese cuisine over the years and highlights its rich heritage and adaptability.

一层平面图
FIRST FLOOR PLAN

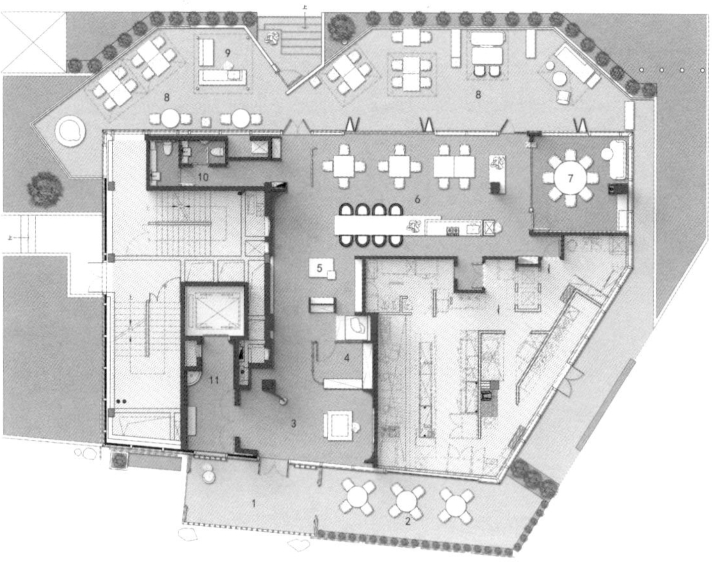

1. 入口花园	ENTRY GARDEN
2. 户外休闲	OUTDOOR RECREATION
3. 大厅	HALL
4. 酒窖	WINE CELLAR
5. 茶艺展示	TEA DISPLAY
6. 散座区	SCATTERED SEATS AREA
7. 包房 1	PRIVATE ROOM 1
8. 户外露台	OUTDOOR TERRACE
9. 酒吧	BAR
10. 卫生间	TOILET
11. 电梯厅	ELEVATOR HALL

二层平面图
SECOND FLOOR PLAN

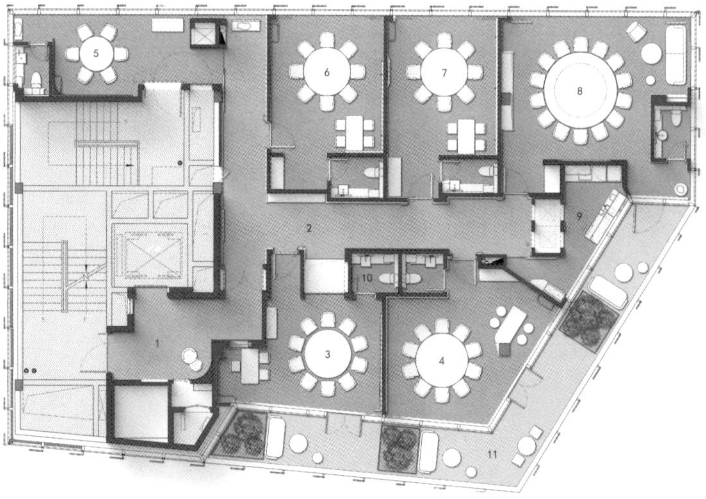

1. 电梯厅	ELEVATOR HALL
2. 卫生间	TOILET
3. 包房 2	PRIVATE ROOM 2
4. 包房 3	PRIVATE ROOM 3
5. 包房 4	PRIVATE ROOM 4
6. 包房 5	PRIVATE ROOM 5
7. 包房 6	PRIVATE ROOM 6
8. 包房 7	PRIVATE ROOM 7
9. 包房 8	PRIVATE ROOM 8
10. 备餐区	PANTRY
11. 室外露台	OUTDOOR TERRACE

舞狮、花灯和木雕是闽南文化不可或缺的一部分，现代与传统共生，时尚与复古俱在。闽南文化在民间工艺、饮食、建筑、南戏等诸多方面，都有传世的非凡发展和创造。

Lion dance, lanterns, and wood carving are integral to Minnan culture, representing a harmonious blend of modernity and tradition, as well as fashion and retro styles. This culture encompasses a rich heritage of folk crafts, cuisine, architecture, and Southern Opera, showcasing numerous remarkable developments and creations passed down through generations.

繁花野趣
FLOURISHING FLOWERS AND WILD CHARM

主题包间的符号性设计，聚焦福建闽南地区的文化特点，包间内百叶屏风优雅而立，拼花地砖绚丽精美，石雕艺术巧夺天工，尽显地域文化之美。

The symbolic design of the theme private room focuses on the cultural characteristics of the Minnan area in Fujian. In the private rooms, the louver screen stands gracefully, the parquet floor tiles are gorgeous and exquisite, and the stone carving art is wonderful, showing the beauty of regional culture blending.

民俗风情

FOLK CUSTOMS

无锡 · 松果果餐厅
SONG GUOGUO

AN ART FEAST AT THE FOOT OF THE MOUNTAIN IN WUXI
独属于无锡山脚下的浪漫江南

项目地点 : 无锡　面积 : 1200 m^2

Location：Wuxi　Area : 1200 m^2

RESTAURANT
SONGGUOGUO

松果果
SONGGUOGUO

松果果的空间设计，以充满东方仪式感的门厅步入方式，以及多层次的变化，体现接待客户的仪式感。一层门厅的旋转楼梯，连接一层和二层空间，楼梯的线条在灯光中被拉长、扭曲、叠加，创造出一种超乎寻常的空间感，让人不禁想要深入其中，一窥究竟。

The spatial design of Songguoguo embodies a sense of ritual in welcoming customers through the entrance approach of Eastern ritual and multi-level changes. The spiral staircase in the lobby on the first floor connects the spaces on the first and second floors. With the lines of the stairs stretched, twisted, and overlaid in a play of light and shadow, it creates an intriguing sense of depth and spatial complexity, inviting people to explore deeper and get a glimplse of it.

一层平面图
FIRST FLOOR PLAN

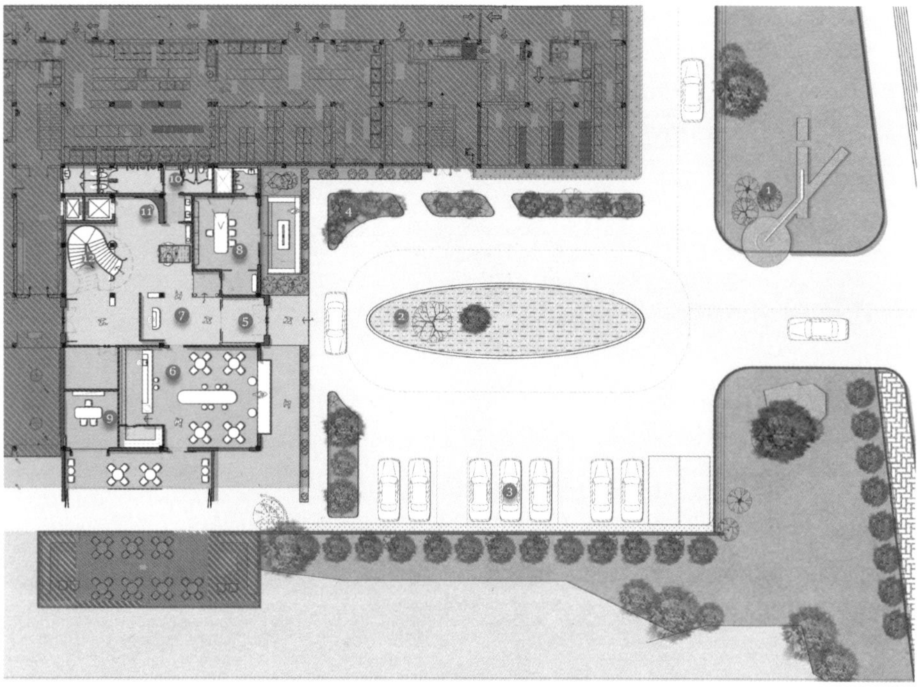

1. 入口装置 ENTRANCE DEVICE
2. 水景 WATERSCAPE
3. 停车区 PARKING AREA
4. 景观池 LANDSCAPE POND
5. 前厅 LOBBY
6. 咖啡吧 COFFEE SHOP
7. 大厅 THE HALL AREA
8. 茶室 1 TEA ROOM 1
9. 茶室 2 TEA ROOM 2
10. 卫生间 BATHROOM
11. 电梯厅 ELEVATOR HALL
12. 楼梯间 STAIRWELL AREA

二层平面图
SECOND FLOOR PLAN

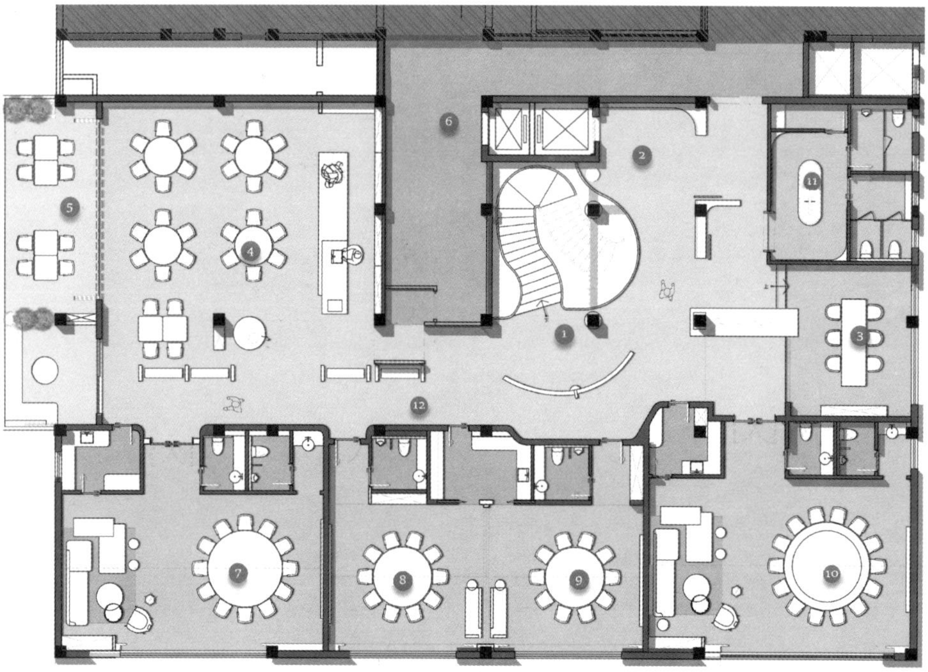

1. 楼梯间 STAIRWELL AREA
2. 电梯厅 ELEVATOR HALL
3. 茶空间 TEA DISTRICT
4. 散座区 SCATTERED SEATS AREA
5. 外摆区 OUTDOOR HALL AREA
6. 厨房后勤 KITCHEN SUPPORT
7. 包间 1 PRIVATE ROOM 1
8. 包间 2 PRIVATE ROOM 2
9. 包间 3 PRIVATE ROOM 3
10. 包间 4 PRIVATE ROOM 4
11. 卫生间 BATHROOM
12. 走廊 CORRIDOR

松果果将作为艺术装置的松树与自然的形貌进行融合设计，营造景观空间的细节与氛围。无论是硬装上的设计元素，还是软装上的隐喻手法，都充分展现了松果果美学的苍劲之感，令人神往。

Songguoguo integrated pine trees as an art installation with natural forms to create details and atmosphere in the landscape space. The design skillfully uses design elements to embody the rich symbolic significance of plants in ancient Chinese culture. The pine tree, with its connotation of tenacity and lasting vitality, is exceptionally well-represented, demonstrating its profound meaning through the design.

逶迤延绵

CONTINUOUS WITHOUT END

二层用餐区的格局设计和多媒体弧幕的设置，打破了空间的视觉界限，构建了一种具有沉浸感的氛围，让人体会到空间与时间交织的韵律。

The design of the dining layout on the second floor and the implantation of the multimedia arc screen break the visual boundaries of the space, and create an immersive atmosphere, allowing people to experience the interweaving rhythm of space and time.

浓郁婉转

RICH AND GRACEFUL

大自然与美食天生带有超强的治愈能量，在松果果享受美食带来愉悦的同时，让时间、文化在精神层面变幻出绵延不绝的余味。

Nature and delicious food are naturally endowed with super strong healing energy. While enjoying the pleasure brought by delicious food at Songguoguo, the time and culture collide to transform an endless aftertaste at the spiritual level.

品牌
BRAND

九储堂
THE NINE

鲁采

金汇酒楼
JINHUI RESTAURANT

永隆龍泰
DRAGON SAWASDEE

JADE VESSEL
鼎琪

PHOENIX39 法餐

YU HUAN
SEAFOOD HOT POT
余欢

莱安五号
LAI'AN NO.5

DIANE FINE CUISINE
黛 外滩

蟹仙画宴
CRAB & PAINTING

文杏馆·海
MAISON GINKGO

ORDER BLESSING
RESTAURANT
福序

SONGGUOGUO
RESTAURANT

KEEY
上海企一

FIVE LISHANG

总策划人：曾建龙

特邀平面设计：陈云飞

灯光设计顾问：朱海燕

灯光品牌：企一照明

摄影师（按姓名拼音排序）：陈荣坤 陈逸飞 李开建 王厅 赵易宏 朱迪 朱海

官网：www.giddesign.com.cn

GID

DESIGN